STEM ABC's
R is for Robot

A child's first ABC book
of Science and Technology

Techie Baby

R is for Robot
Written by Sheyna Sears

ISBN-13: 978-1-7322633-0-7

Published by Techie Baby Books, LLC
Chapel Hill, North Carolina, USA

Dear Parents, Teachers, and Caregivers,

The modern baby is growing up in a society driven by science, technology, engineering, and math (STEM). Early exposure to STEM concepts will be essential for our children to have the greatest success in tomorrow's world.

With Techie Baby books, you begin introducing STEM ideas from the very beginning so that your little ones are never intimidated by math or science. They will approach such concepts with the same confidence and enthusiasm that they have when making animal sounds or naming colors. As your child sits on your lap and enjoys the rhymes and pictures in this book, you are helping your little one develop scientific intuition. That intuition, combined with a child's natural playfulness, is what leads to the technological breakthroughs that change our world.

You will be amazed by how quickly your readers will begin to make connections with the world around them - noticing a pulley on a flagpole, observing that the ball rolls farther on the sidewalk than on the grass, asking how and why.

And so, the young child enters the world of technology and innovation and begins a lifetime of learning and discovery.

With love and excitement for the future,

Sheyna Sears

A is for Astronaut exploring space.

B is for Building a brand new place.

C is for Compass to show us the way.

D is for Discovering new things each day.

E is for Electricity to light up the town.

F is for Fossils
buried deep down.

G is for Gears spinning around.

H is for Horizon
where the sky meets the ground.

I is for Invent - create something new!

J is for Jet flying into the blue.

K is for Kindness.
Use what you learn to do what is right.

L is for Lightning
that makes the sky bright.

M is for Measuring how long or how tall.

N is for Nature. Take care of it all.

O is for Orbit -
the path of a planet moving around.

P is for Pulley - to lift heavy objects high off the ground.

Q is for Question - wonder and ask!

R is for Robot to help with our tasks.

S is for Scientist who figures out why.

T is for Telescope to look at the sky.

U is for Unknown.
There's so much to explore.

V is for Volcano
with hot rocks from Earth's core.

W is for Wheel to roll and to spin.

X is for X-ray. See bones deep within.

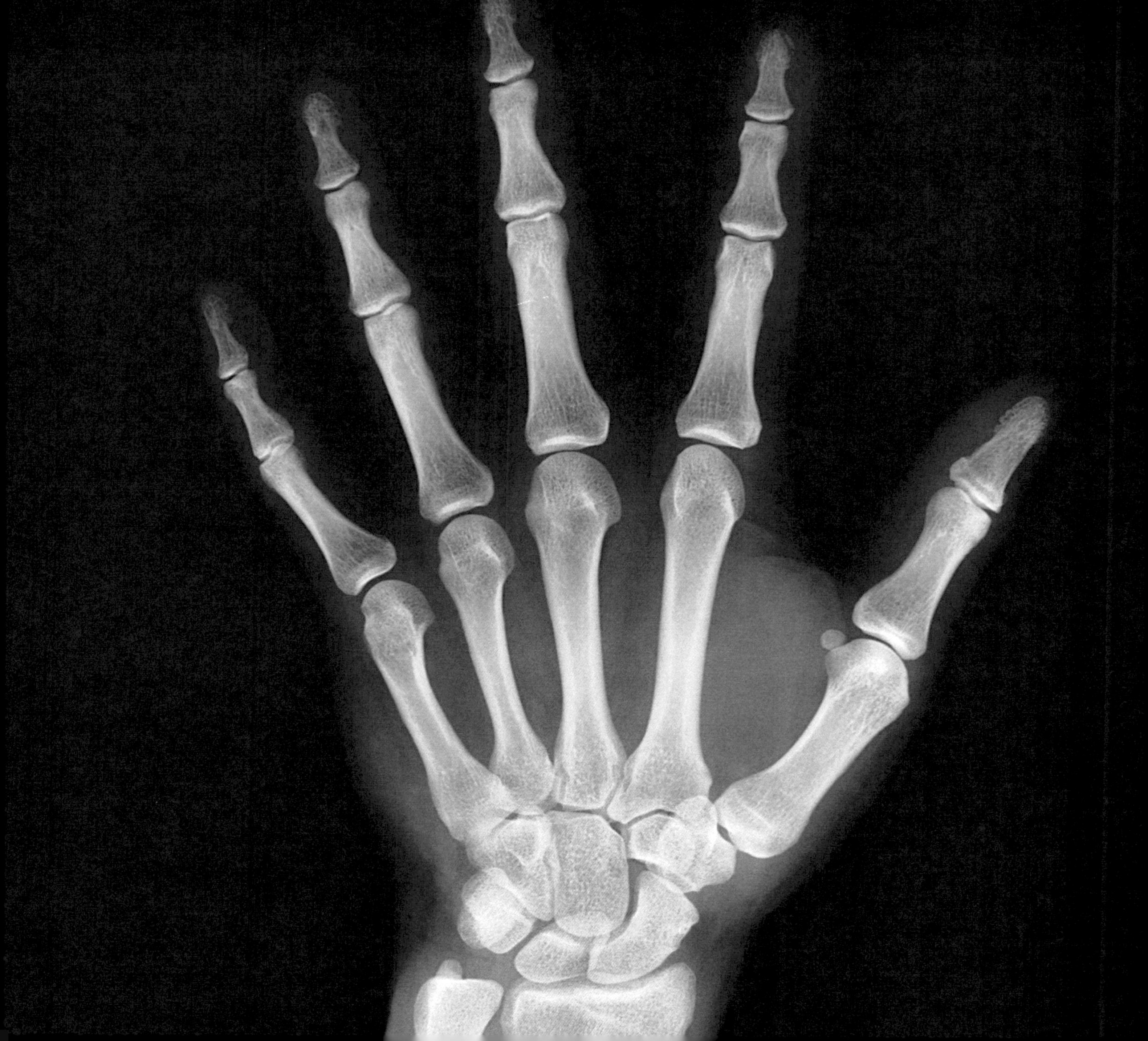

Y is for Year. One trip around the sun.

Z is for Zany. Think outside the box and have lots of fun!

Glossary

More information to help answer children's questions and encourage further discussion.

Astronaut – An astronaut is a person who travels in a spacecraft beyond earth's atmosphere. The astronaut in the photo is outside his shuttle and is wearing a special suit for protection.

Building – Many people work together to make a building. Some people design how the building should be put together so that it is strong and stable, and other people work to put the pieces of the building together using machines such as bulldozers, diggers, and cranes.

Compass – A compass helps us figure out direction so we don't get lost. A magnet inside the compass points the arrow towards the north pole.

Discovery – Discovering something is finding or realizing something new. Discovery is exciting because it includes exploring, observing the world, trying new things, and asking questions.

Electricity – Electricity is energy from tiny particles called electrons moving around.

Fossil – Fossils are parts of ancient living things that have been preserved in mud. Paleontologists study fossils of animal bones, shells, leaves, and even footprints to learn more about the types of animals and plants that used to be here.

Gears – Gears look like wheels with teeth. The teeth or bumps on one gear can connect to the bumps on another gear. When the gears rotate together, they can change the speed or direction that something is moving.

Horizon – The horizon looks like a line separating the ground or sea from the sky. It can be hard to see the horizon if trees or buildings are in the way.

I**nvent** – Inventing is creating something that wasn't there before. Toys and cars and even cookies didn't exist until someone invented them. You can create something new by paying attention to what people need and by thinking of a new way to do something.

Jet – A jet plane uses an engine that mixes air and fuel to push the plane forward. If you blow up a balloon and let it go, the air in the balloon will blow in one direction, making the balloon fly in the opposite direction - sort of like a jet.

Kindness – Kindness is being nice and helpful and thinking about what other people want.

Lightning – Lightning is a form of electricity that occurs during thunderstorms. Little particles, called electrons, move around during the storm and create a bright light, called lightning.

Measure – Measuring is figuring out the size or amount of something. We can use different measuring tools to learn how tall, how fast, how long, how heavy, or how hot something is.

Nature – Nature includes everything in our world that people did not create. Rocks, trees, water, clouds, and animals are all part of nature.

Orbit – An orbit is a path shaped like a circle or an oval. The planets orbit around the sun.

Pulley – A pulley is a type of machine with a rope or belt that moves around a wheel. Pulleys can help lift heavy weights and change the direction of a force (e.g., a person pulls down to make something move up). There are pulleys in many curtains and flagpoles.

Question – Questions help us learn and understand new things. You can ask, What is something? How does it work? Why is it useful? or anything else you are wondering.

Robot – A robot is a machine that can do a complicated task, like building a car or cleaning the floor in your house. A person can use a computer to program the robot and make it work.

Scientist – A scientist is someone who observes the world, asks questions, and tries to figure out how things work.

Telescope – A telescope makes things that are very far away look bigger so that we can see them better.

Unknown – If something is unknown, it means we don't know much about it. The unknown is exciting because it gives us a chance to learn and explore.

Volcano – A volcano is a type of mountain. Steam, ash, and hot rocks, called lava, can explode out of a hole at the top of the volcano.

Wheel – A wheel is shaped like a circle and spins around an axle at its center. Wheels help things roll and move more easily over the ground. Bikes, cars, and wagons all have wheels.

X-ray – An x-ray is a type of picture of the inside of your body. We can use x-rays to see our bones and make sure they are healthy.

Year – A year is 365 days. It takes one year for our earth to move all the way around the sun. In that time we have spring, summer, winter, and fall, and we grow one year older.

Zany – Being zany is being silly and fun. Have fun as you learn about our world and come up with your own amazing ideas and questions.

Photo Credits

Thank you to all of the designers, photographers, and models who make the world a more beautiful and interesting place. *R is for Robot* and Techie Baby Books, LLC have not been authorized, sponsored, or otherwise approved by any of the products or companies depicted in this book.

Astronaut	Photo by NASA via Wikipedia Commons
Building	Photo by MichaelGaida on Pixabay
Compass	Photo by StockSnap on Pixabay
Discovery	Photo by MI PHAM on Unsplash
Electicity	Photo by Zoltan Kovacs on Unsplash
Fossil	Photo by kkorvin on Pixabay
Gears	Photo by fsHH on Pixabay
Horizon	Photo by Rodion Kustaev on Unsplash
Invent	Photo by LisaRedfern on Pixabay
Jet	Photo by Jorden Sanchez on Unsplash
Kindness	Photo by Ben White on Unsplash
Lightning	Photo by Jeremy Bishop on Unsplash
Measure	Photo by Stevepb on Pixabay
Nature	Photo by Christian Joudrey (@cjoudrey)on Unsplash
Orbit	Picture by NASA.gov PIA12114_0
Pulley	Photo by ranjatm on Pixabay
Question	Photo by Les Anderson on Unsplash
Robot	Photo by Alex Knight on Unsplash
Scientist	Photo by LinkedIn Sales Navigator on Unsplash
Telescope	Photo by Teddy Kelley on Unsplash
Unknown	Photo by Ksenia Makagonova on Unsplash
Volcano lava	Photo by orcrist on Pixabay
Wheels	Photo by Alexander Shustov on Unsplash
X-ray	Photo by com329329 on Pixabay
Year	Photo by m_u_z007 on Pixabay
Zany	Photo by Porapak Apichodilok on Pexels

Cover Photos

Astronaut	Photo by NASA via Wikipedia Commons
Volcano	Photo by usgs.gov img5123
Gears	Photo by Pexels on Pixabay
Robot	Photo by Alex Knight on Unsplash

Made in the USA
Las Vegas, NV
07 November 2026

57672512R00021